AF706005

The AI Culture Blueprint

Moving Beyond Tools to
Create Human-Centered AI Adoption

Sagar Pandya

Copyright © 2025 by Sagar Pandya

All rights reserved.

No portion of this book may be reproduced in any form without written permission from the publisher or author, except as permitted by U.S. copyright law.

This publication is designed to provide accurate and authoritative information in regard to the subject matter covered. It is sold with the understanding that neither the author nor the publisher is engaged in rendering legal, investment, accounting, health, or other professional services. While the publisher and author have used their best efforts in preparing this book, they make no representations or warranties with respect to the accuracy or completeness of the contents of this book and specifically disclaim any implied warranties of merchantability or fitness for a particular purpose. No warranty may be created or extended by sales representatives or written sales materials. The advice and strategies contained herein may not be suitable for your situation. You should consult with a professional when appropriate. Neither the publisher nor the author shall be liable for any loss of profit or any other commercial or damages, including but not limited to special, incidental, consequential, personal, or other damages.

ISBN: 978-1-968247-37-9 (Digital)

ISBN: 978-1-968247-38-6 (Paperback)

Book Cover by 100Covers

Edited by Bailey Baack

1st Edition 2025

About the Author

Sagar Pandya is a serial entrepreneur, AI strategist, culture consultant, and the founder of The AI Culture Company, where he helps entrepreneurs and business leaders navigate how artificial intelligence is reshaping the way work actually gets done. With over two decades of experience across IT, cybersecurity, and digital transformation; and firsthand experience leading a fifty-person AI consulting organization following an acquisition; he has seen why so many AI initiatives stall despite significant investment. Through speaking, writing, and advising, Sagar focuses on the human side of AI adoption, helping organizations move past fear and uncertainty to build AI-ready cultures where teams of people are involved, informed, and empowered alongside the technology.

Contents

INTRODUCTION

The AI Culture Lag Crisis

We live in a moment where artificial intelligence dominates nearly every conversation. Media coverage, social posts, curiosity, and fear have pushed it to the center of global attention, and for good reason. AI is advancing at lightning speed and most organizations are struggling to keep up.

The instinctive response in the business world has been to buy new tools and hope adoption follows. But AI represents a fundamental shift in how we work. Too often, companies focus on the software while neglecting the culture that makes adoption possible. The result is stalled projects, wasted investment, and mounting frustration.

AI doesn't fail because of the tools. It fails because of ignored human dynamics. Many consultants and speakers focus on features, platforms, and capabilities, assuming everyone from interns to executives will naturally adapt. But they overlook the human experience and the identity crisis created when a machine can replicate decades of expertise in seconds. They overlook the discomfort of experimenting publicly and the fear of failure. Unless we address these realities, AI will remain blocked, bottlenecked, and underused.

A change of this magnitude demands cultural readiness. When a forty-year veteran is told that a machine can complete their job in minutes, this causes a profound shift in identity. If organizations don't navigate this human side of adoption, they will lose ground. Customers will drift, talent will leave, and competitive advantage will erode. When we approach AI from the bottom-up, treating it as a people-first transformation, we see adoption accelerate and teams engage.

The speed of AI's evolution and the tendency to focus on tools instead of culture makes this a genuine crisis. Companies that adapt their culture to move as quickly as technology will thrive. Those that don't will fall behind. Talent, revenue, and entire industries are at stake. Bringing AI into the cultural fabric of organizations may be the most important leadership challenge of the next decade.Whether your AI investment is $50,000 or $5 million, the culture dynamics remain the same.

This book is designed to help you meet that challenge. Whether you're a founder, c-suite executive, leader, manager, or individual contributor, you'll find practical steps and reflective questions to guide your journey. Are AI conversations happening in your workplace? Do your people feel safe sharing both fears and experiments? Do you reward curiosity, or do you unintentionally silence it?

The decade ahead will be defined by how organizations answer these questions. Every company, whether a startup or a global enterprise, now has access to the same AI capabilities. The differentiator will be culture.

Through the 5C Framework for AI Culture—Commit, Communicate, Co-Create, Coach, and Cultivate—this book gives you a blueprint to start small, build momentum, and sustain real change. Each chapter ends with practical tools you can use immediately in your team or organization. By the end, you'll not only have the foundation of your AI strategy, but also the confidence to lead with curiosity, resilience, and clarity in a world being reshaped by artificial intelligence.

Action Tool #1:
The AI Culture Self-Assessment Tool

Directions: Reflect on this critical question - Are you crawling, walking, or running with AI? What's holding you back from the next stage? This assessment determines your current AI maturity and identifies barriers to advancement. Use this assessment to evaluate your organization's current AI readiness.

Rate each statement from 1 (Disagree) to 3 (Neutral) to 5 (Agree):

______ Team members feel comfortable sharing AI experiments and failures.
______ We address concerns about job displacement openly and honestly.
______ We measure success beyond traditional metrics.
______ I can clearly articulate in thirty seconds why our organization needs to embrace AI.
______ Our team knows whether I'm excited or skeptical about AI (I've been transparent about my perspective).
______ We allocate a meaningful budget to AI culture change, not just technology tools.
______ I know which team members are already using AI tools (even if we haven't officially approved them).
______ Team members feel safe admitting they've tried AI tools or sharing when experiments fail.
______ I regularly ask my team about their AI fears, excitement, and ideas (not just in formal meetings)
______ When we select new tools or processes, we include frontline employees in the decision, not just leadership.
______ Our organization celebrates both successful experiments AND intelligent failures.
______ We have people from different departments and experience levels working together on AI initiatives.
______ Our managers know how to coach team members who are skeptical about AI.
______ People share AI wins and learnings across departments (not just within their own teams).
______ We measure AI adoption success by engagement and behavior change, not just ROI.
______ Our team sees AI as enhancing their work rather than threatening their jobs.
______ We have clear guidelines about what AI experimentation is encouraged vs. discouraged.
______ AI conversations happen regularly in our organization without needing to be forced or scheduled.

Scoring Your AI Culture Stage:

______ / **90**

66-90 = Running with AI
43-65 = Walking with AI
18-42 = Crawling with AI

CRAWLING (18-42 points)	WALKING (43-65 points)	RUNNING (66-90 points)
Where you are: AI is new, scary, or confusing for your organization. Most conversations happen at the leadership level, and there's significant uncertainty about how to move forward. **What this means:** You're in the foundation stage. Focus on building trust, understanding current state, and having honest conversations before investing in tools. **Your next step:** Complete your "why" statement and start listening tours with your team. Don't buy AI tools yet, but build the culture first.	**Where you are:** You have some momentum and early adopters, but adoption is inconsistent across the organization. Some people are excited, others are hesitant or resistant. **What this means:** You're in the scaling stage. You have proof of concept but need better systems to spread success and address resistance. **Your next step:** Form your AI Council and create systematic ways to share learnings. Focus on peer-to-peer coaching and celebrating both wins and failures.	**Where you are:** AI adoption feels natural and organic in your organization. People proactively suggest improvements, failures are treated as learning, and AI enhances rather than threatens your culture. **What this means:** You're in the optimization stage. Focus on measurement, sustainability, and staying ahead of the curve as AI evolves. **Your next step:** Develop sophisticated measurement systems and prepare to help other organizations learn from your success.

You can also access this tool online at **aicultureblueprint.com/culturesurvey**.

CHAPTER ONE

Commit: *Lead with Why*

When discussing culture and change with organizations, the first question I ask leaders is, *"Why?"* Why are we deciding to go down this path right now? Why wasn't this considered three months or three years ago? Why aren't we waiting another three years?

I learned to start with "why" the hard way earlier this year. A 200-person professional services firm called me after spending $50,000 on AI tools that nobody was using. The CEO was frustrated: *"We bought the best technology, sent the announcement email, even did training sessions. But six months later, adoption is at 12 percent. What are we doing wrong?"*

Leading with "why" is fundamental. The sooner you cultivate and create your "why" statement, the sooner you'll have a north star guiding every AI-related decision moving forward. When leaders can articulate their "why' to their teams while giving teams the ability to contribute, it creates a culture of shared responsibility.

We cannot make leadership decisions from C-suite towers and expect teams to follow every path we build. The bottom-up approach remains the only way to improve AI culture within organizations.

Real-World Examples of Why

Across industries, I've heard many different "why" statements from leaders. Some companies say, *"We don't have the resources to hire more people, but we need to keep supporting our clients."* In today's tight talent market, finding employees who align with company culture is increasingly difficult. For these organizations, the motivation is to relieve the pressure on their teams while continuing to deliver value.

Others take a different view: *"We're not adding headcount. We already have a strong team and want to elevate them. But as a publicly traded company, we're accountable to shareholders. We have to hit quarterly numbers in a competitive market."* For them, the "why" is profitability. Without revenue, nothing else—staff, AI tools, or culture-building—can be sustained.

Some organizations wrestle with true talent shortages. Even with the resources to hire, they can't find qualified candidates. In certain industries, AI can fill these gaps by handling tasks more efficiently than someone with entry-level training.

Then there are the innovators. These companies operate in fast-moving fields like biotech or agriculture, where research and development depends on processing massive datasets quickly. For them, AI is less about filling gaps and more about fueling breakthroughs.

The specific "why" varies, but the principle stays the same: Define it, live it daily, and invite your teams to contribute. A clear "why" creates alignment and purpose, but it must be shared, not dictated.

Leaders don't have to carry this burden alone. Businesses evolve far beyond the vision of their creators, shaped by the people who now run them. You may no longer see customers every day or fully grasp how your company's DNA has shifted over time. That's why empowering teams to help shape the "why" is essential. Their voices will keep you on course and strengthen the north star that guides your organization forward.

Budget as Commitment Signals

Don't forget, people watch where you spend money, not where you say your priorities are. I've seen this repeatedly with AI implementations. A CEO will stand in front of their team and talk passionately about embracing AI culture, building trust, and empowering people to experiment. Then they allocate $500,000 to technology tools and $30,000 to training. What message does that send?

Your team isn't stupid. They can do math. When 94 percent of your AI budget goes to technology and 6 percent goes to people, they understand exactly what you actually value. You can talk about culture all you want, but your budget tells the real story.

This creates a credibility gap that kills AI adoption before it starts. Teams think, *"They say they care about our input, but they've already bought all the tools. They say they want us to feel comfortable experimenting, but they're not investing in helping us learn."*

The companies I work with that see real AI adoption typically flip this ratio. They spend 60 percent on culture with listening sessions, training, change management, coaching, and 40 percent on technology. This sounds backward to finance people who want to see immediate returns on software purchases. But when you invest in culture first, the technology actually gets used. When you invest in technology first, you get expensive shelf-ware.

I worked with a manufacturing company that spent six months and $775,000 on an AI-powered inventory management system. Eighteen months later, adoption was at 15 percent. The operations manager told me, *"The system works fine. People just don't trust it, and we never gave them time to understand how it could help them."* They had to spend another $240,000 on changing management and training to get adoption above 70 percent.

Your budget allocation is your real commitment statement to support your "why." If you're serious about AI culture, put your money where your mouth is. Invest in the people who have to make the technology successful, not just the technology itself.

The beauty of the "why" statement lies in its simplicity and authenticity. Don't overthink this process or fall into the trap of endless revision cycles with analysis paralysis. Your initial "why" statement is a living document that will evolve as your organization grows and learns. The key is to begin the conversation and establish the north star immediately.

When you sit down with pen and paper, focus on what genuinely drives your decision to embrace AI today. Is it competitive pressure? Resource constraints? Innovation opportunities? Growth ambitions? Whatever surfaces in those first thirty minutes of honest reflection is your starting point. Your team will appreciate the transparency and authenticity far more than a perfectly-crafted statement that took weeks to develop but feels disconnected from reality. This "why" statement becomes the foundation for every AI decision, every cultural shift, and every strategic conversation moving forward. Start today, refine tomorrow, but most importantly: **Start.**

Reflection Questions

Before using the AI Culture Commitment Builder tool below, examine your own mindset and readiness. These questions are internal checks to ensure you approach the tool with the right perspective.

Leadership Mindset Check: Are you clearly articulating the why because you genuinely believe it, or because you think it's what you should say? Your team will sense the difference between authentic commitment and corporate speak. If you're afraid to articulate your real reasoning because you're unsure of your team's reaction, that's a red flag worth exploring.

Your AI Perspective: What do you personally think about AI? As leaders, you model behavior that your team will follow. If you're a skeptic waiting for the technology to mature or a perfectionist who can't tolerate experimental failures, these perspectives will unconsciously influence your approach. Before you realize it, you become the bottleneck preventing your team from trying new approaches.

Vulnerability Factor: Are you willing to share your own AI learning journey, including uncertainties and mistakes? People embrace vulnerability and transparency in organizations. This becomes the glue holding together the culture you're building.

If you find yourself deflecting these internal questions, spend more time in honest reflection before moving on. A confused or inauthentic "why" leads to confused execution and wasted investment.

Action Tool #2:

The AI Culture Commitment Builder

pg. 1

Step 1: What is driving your AI decisions right now?

Circle the letter that best describes your situation:

A) Resource Constraints — "We don't have enough people to handle our workload."
B) Competitive Pressure — "Competitors are moving faster than us."
C) Efficiency/Margin Pressure — "We need to improve profitability without increasing costs."
D) Innovation Ambition — "We want to lead our industry with cutting-edge capabilities."
E) Talent/Skill Gaps — "We can't find or keep the right people."
F) Customer Demands — "Customers expect faster, smarter service."

Step 2: How urgent is this?

Check one:

A) Critical (next 30 days) — "We're already falling behind."
B) Important (next 6 months) — "We need to start soon."
C) Strategic (next year) — "Positioning for the future."

Step 3: Budget Reality Check

What percentage of your AI budget goes to each area?

Culture & Training: _____ %

Technology & Tools: _____ %

(Must total 100%)

Step 4: Write your "why" statement.

Based on your Step 1 selection, complete this sentence:

If you selected A: "We're embracing AI because we need to serve our customers without burning out our team or compromising quality."
If you selected B: "We're embracing AI because staying competitive means moving faster than companies still doing everything manually."
If you selected C: "We're embracing AI because sustainable profitability requires us to deliver the same quality more efficiently."
If you selected D: "We're embracing AI because leading our industry means creating capabilities others can't match."
If you selected E: "We're embracing AI because building a great team means attracting and enhancing forward-thinking people."
If you selected F: "We're embracing AI because our customers deserve faster, smarter service than manual processes can provide."

Write your personalized "why" statement below:

Action Tool #2:
The AI Culture Commitment Builder

pg. 2

Step 5: Your Leadership Commitment	**Step 6:** Validation Check
Complete this statement: "As a leader, I commit to ____________________ ______________________________ to ensure our AI adoption succeeds." Examples: • "allocating 30 minutes weekly to AI culture conversations" • "being transparent about my own AI learning journey" • "measuring success by team adoption, not just ROI" • "investing in people development alongside technology"	Before sharing this with anyone, ask yourself: • Can I say this "why" statement in 30 seconds without notes? • Does this sound authentic to how I normally communicate? • Would I base budget decisions on this reasoning? • Am I ready to have honest conversations with my team about this?

Your Next Steps

This week:

- Share your "why" statement with one trusted advisor
- Ask for their honest feedback on authenticity and clarity

Next 30 days:

- Use this "why" to evaluate any AI opportunities that arise
- Begin planning how you'll share this with your team

You're ready for Chapter Two when you can confidently affirm all validation criteria above. Complete an enhanced version of this tool online at: **aicultureblueprint.com/whybuilder.**

CHAPTER TWO

Communicate: *Listen Before You Launch*

This chapter matters deeply because it puts listening at the center of AI adoption. Leaders often believe they already know what their teams need automatically, but the truth is the answers are usually right in front of us if we take the time to listen.

Too often, executives make assumptions, roll out expensive tools, and then feel blindsided when adoption lags. It's the classic "ready, fire, aim" mistake: launching solutions before understanding the real problems.

I saw this firsthand with a 150-person marketing agency. Their CEO had invested $153,000 in an AI content platform, only to watch it sit unused for months. When I asked if she had spoken with her team about their needs, she explained that she had sent a survey, but only after purchasing the tool. What she hadn't realized was that her creative team had already been experimenting with free AI tools like ChatGPT and Claude for months. They had built workflows, refined prompts, and improved efficiency by nearly 30 percent. What they needed wasn't a new platform. It was training to deepen their skills and support integrating AI into their existing processes.

When leaders skip listening, they risk solving the wrong problems. Communication in this context means creating ongoing conversations across every level of the organization.

Listening to fears, capturing ideas, and surfacing grassroots innovation lays the groundwork for meaningful adoption.

Starting the Conversation

This conversation can begin anywhere, whether with senior leaders, frontline employees, or interns. Every voice matters, and each group brings a unique perspective. Senior leaders understand strategic pressures, frontline workers see daily operational challenges, and newer employees often question long-standing assumptions.

The challenge is that many people won't speak up without a sense of safety. That safety isn't created by sending out a quick survey link. It's built through multiple channels that fit different personalities. Extroverts may prefer group discussions, introverts often thrive in one-on-one settings, and some people will only contribute anonymously. Offering all of these options ensures that valuable perspectives aren't lost.

The goal is not just to collect comments, but to genuinely listen, placing yourself in someone else's shoes and understanding how they experience change. When leaders create these kinds of safe, flexible spaces, they uncover the insights needed to guide adoption in a meaningful way.

The Three-Channel Approach

Group settings work for extroverted people and verbal processors who thrive when sharing with peers. They need platforms to think out loud and build energy from others' ideas.

One-to-one conversations serve introverted folks who don't want to be on stage or put on the spot. They need the questions beforehand and time to think before they can deliver thoughtful insights.

Anonymous feedback tools capture input from people who won't share any other way. This still develops your communication and listening skills to build that north star for AI culture.

Through these conversations, you will discover which team members are already experimenting with AI tools, what specific fears are preventing adoption, which processes are crying out for automation, and what support people need to feel confident moving forward.

Using AI transcription tools becomes valuable here. AI can objectively capture conversations to review later, articulating and highlighting critical points for culture development. These tools deliver a running list of conversations you can reference without retracing steps or relying on potentially incorrect memory. This approach allows meaningful conversations because instead of spending time taking notes, trying to remember what was said, or keeping up with someone's speaking pace, you can focus on genuine dialogue.

Understanding FUD:

Fear, Uncertainty, and Doubt

The biggest detractor from culture development is not understanding the different levels of fear, uncertainty, and doubt (FUD). Some people in your organization are FUD-free and excited about everything. Others are six feet deep in FUD and need help getting pulled out of that to a level where they can thrive.

Imagine if your company culture was one where people's egos and identities blocked innovation and experimentation. We can't get upset at someone for spending thirty-five years learning a craft and then reacting badly to being told they must outsource it to a machine. We must be sensitive to these topics, giving people chances to explain their feelings. All this information reveals who is using AI and who isn't, their perspectives, and their feelings on it. If they prefer not to use it, what's the reason? Is their identity so attached to their work that any potential threat gets completely shut down?

The Three Levels of FUD

Personal FUD: Job security fears, identity at work, concern about becoming obsolete

Process FUD: Doubt that AI will actually work, skepticism about quality and reliability

Cultural FUD: Belief that this is just another management initiative that will fade away

For larger companies, understanding FUD becomes a tedious task requiring you to arm and elevate leaders in your organization to help with data capture. This is especially important for larger organizations because you lose identity when there are thousands of employees.

When we conduct listening exercises around FUD, our people feel heard and they show up better for customers. Excited customers spend more money, creating a domino effect that unlocks more budget for AI tools. It's a positive cycle where addressing fears leads to better performance, better customer relationships, and ultimately more resources to invest in AI initiatives.

Reflection Questions

1. When was the last time you had a one-on-one conversation with someone three levels below you in the organization? If it's been more than a month, you're probably not as connected to your team's reality as you think you are.

2. Are people truthful about their perspectives, or telling you what they think you want to hear? If you're not sure, that's probably your answer. People who feel safe to disagree will tell you directly.

3. Is there experimentation happening that people are afraid to share because your culture doesn't praise mistakes or vulnerability? This is where you'll find your biggest AI opportunities and your biggest cultural barriers.

4. Think about the last three times you asked your team for feedback on anything. What did you do with that feedback? If you can't remember specific actions you took, why should they bother sharing their AI perspectives with you?

5. Are there departmental silos preventing you from discovering innovation that's already happening?

6. Does your organization collaborate and celebrate wins and losses together? How do people typically react when someone admits they tried something that didn't work?

Your answer determines whether people will share their real AI experiences with you.

7. What decisions about AI tools or strategy have you made in the last six months without asking your team for input first?

Action Tool #3:

The AI Listening Tour Toolkit

Note: This tool mentions the AI Council, which will be covered in the next chapter.

Step One: Who To Talk With

Target eight to twelve conversations across:

- Senior leaders
- Middle managers
- Frontline workers
- New employees
- Veterans (10+ years)
- Tech-savvy
- Tech-hesitant

Step Two: Three Question Sets

GROUP CONVERSATIONS (3-8 people):

- What's your current experience with AI tools?
- What excites you about AI possibilities?
- What concerns do you have?
- Where do you see the biggest opportunities?

ONE-ON-ONE CONVERSATIONS (send questions ahead):

- Walk me through your typical day. Where do you get stuck?
- What repetitive tasks eat up your time?
- Have you experimented with AI tools?
- What would make you excited vs. worried about AI at work?

ANONYMOUS SURVEY:

- Rate your AI comfort level (1-10)
- What's your biggest AI concern?
- What AI experimentation haven't you shared?
- What would make you feel safe trying AI tools?

Step Three: Simple Capture Method

For each conversation, note:

Person: ______________

Date: ______________

Current AI usage: ________

____

Biggest excitement: ____

____

Primary fear: ________

____

Fear, Uncertainty, and Doubt (FUD) Level:
☐ Excited ☐ Curious
☐ Skeptical ☐ Resistant

AI Council potential: ☐ Yes ☐ Maybe ☐ No

Step Four: What You're Looking For

Use AI transcription tools to capture conversations so you can focus on listening instead of note-taking. You're ready for Chapter Three when:

- You've talked to people at different levels
- You know your team's top 3 AI opportunities and concerns
- You've identified potential AI Council members
- You understand what support people need

Complete an interactive version at:
aicultureblueprint.com/listeningtour

CHAPTER THREE

Co-Create: *Design Together, Adopt Faster*

The concept of AI Councils inside organizations has become a powerful tool in business. I saw a different approach fail spectacularly this year with a 900-person logistics company. The CEO and CTO spent six months evaluating AI platforms, chose an expensive workflow automation system, then announced the rollout to department heads. Six months later, the system was barely used. When I asked why, the operations manager said, *"It doesn't work the way we actually do things. If they'd asked us, we could have told them that our processes are completely different from what they assumed."*

The IT department had to spend another four months customizing the system to match real workflows. This is work that could have been avoided entirely if they'd included operational people in the selection process from day one. This is exactly why AI Councils with cross-functional and cross-level representation work so much better within your organization when including senior leaders, frontline folks, and everyone in between in the conversation.

The magic happens when you mix these different types: senior leaders who understand strategic pressures, frontline folks who know operational pain points, and newer employees who have fewer preconceptions about "how things have always been done." You'll have folks excited about AI usage, perhaps using it in their personal lives or in their work for years without your awareness. Then you'll have resistant folks who haven't converted yet, which is what this book addresses.

The Right Mix for an AI Council:

- Senior and non-senior leaders
- Tech-savvy and non-tech-savvy
- Tenured and new employees
- Extroverted and introverted
- Excited and resistant to AI

Building the Right Team

An AI Council establishes a powerful strategy by utilizing five to six people. If you need a larger group, choose no more than ten. Too many opinions complicate things. If necessary, break them down by department, business unit, or function. Having a small group meet regularly works best.

If you have a room full of sales folks talking about how AI can transform selling, how does that make finance people feel? Finance people aren't incentivized to sell; they're incentivized for number accuracy. Having broad representation gives you the best opportunity to get everyone's perspective heard and gives people chances to contribute.

When you get a group of individuals from tech-savvy folks to seasoned veterans to senior leaders who make difficult daily decisions and frontline employees who keep businesses running in a room together and talk about AI, there's magic every single time. You already have all the answers within the organization.

Someone in that Council must be the AI Champion, and this choice will make or break your entire AI culture effort. This person becomes the internal energy source driving everything forward, so you can't get this selection wrong. First, it cannot be an external person. Don't rely on external consultants to be your AI Champions because DNA and cultural development must happen from within. It cannot happen with external or fractional people.

I've seen companies try to use their IT vendor, Managed Service Provider, or external AI consultant as the champion, and it always fails because they don't understand the company's real culture and politics. Internal IT people sometimes struggle with this because they have other duties that are more timely and pressing. This becomes another burden on their plate. If you can avoid that, give your IT folks a break from owning the AI conversation.

What Makes an Effective Champion

Your AI Champion must be someone internal, someone good with communication, well-respected in the organization, interested in improving the company's future, incentivized, and driven to move things forward. This allows them to be that internal energy source that keeps momentum alive when initial excitement fades.

Look for someone who naturally connects across departments. They don't need thirty years of AI experience. They need to manage communication and culture. Maybe it's your head of operations whom everyone trusts, or a senior manager whom people actually listen to, not just someone with the right title on an org chart.

Give them whatever title you think appropriate. Some companies are formal and sophisticated, appointing a Chief AI Officer. Others go with a Chief AI Enabler (someone who enables AI usage) or simply an AI Champion. Get creative with titles because these AI Champions will be excited to be given opportunities to innovate. They'll drive and influence the "why" while influencing your ability to collect listening data.

The champion facilitates council meetings, ensures agreements and guidelines get followed, captures intelligence from conversations, and becomes the bridge between what the Council discovers and what leadership needs to know. Without the right person in this role, your Council becomes just another meeting that gradually loses energy and focus. The AI Council in your organization, led by your AI Champion, will be the reason your "why" continues with forward progress.

Making it Sustainable

Having a small group meet regularly works best. Thirty to sixty minutes every one or two weeks provides more than enough time. Do this for six months and you'll gather tremendous amounts of data through regular consistency.

The AI Council in your organization must be one of your company's absolute top priorities. This cannot be an afterthought, cannot be cannibalized, and cannot be delayed because there are bigger business priorities. Of course, there are exceptions, but things shouldn't happen reactively on a regular basis. If they do, I would rethink why that's happening regularly.

Meetings start on time and end on time, with a routine around note collection. People must be willing to show up. If they can't attend, substitute someone from that function so there's continued representation. There must be ambition. The number one thing I think about in AI Councils are folks who are ambitious and want to push the envelope, take charge, and challenge the organizations they're in.

This isn't always the case. Some companies struggle with prioritizing specific topics internally, often due to a lack of maturity around decision-making. But it can be solved through conversations and clear expectations.

From Intelligence to Action

The AI Council doesn't exist just to talk about AI. It exists to drive decisions and energy. The agreed-upon actions and priorities from each AI Council meeting determine the AI tools used in the company, the experimentation that happens, and the testing of productivity improvements. Use the decision-making framework you already operate on within the company. The AI Council should behave like any other team within the company that's accountable for results.

Without this connection to real decisions, your AI Council becomes a discussion group that frustrates ambitious people and wastes everyone's time. The group's intelligence must flow directly into overall company strategy, tool selection, and resource allocation. Otherwise, you're just creating another meeting for people to attend.

Reflection Questions

1. How do you as a company make decisions? (Forget AI—any decision: vendor selection, supplier agreements, customer targeting, hiring)
2. In that decision-making mentality, who is included?
3. Are people who will be directly impacted included in decisions?
4. Are people willing to open up so that when decisions get made, they solve real pain points you may not be aware of?
5. When decisions don't go in the right direction, how do you recover?
6. Who in your organization naturally brings people together and gets things done?
7. When you've formed cross-functional teams before, what made them successful rather than just another meeting?
8. Are you prepared to let this council influence real decisions, or are you just looking for input to validate choices you've already made?

Action Tool #4:
The AI Council Starter Kit

Step One: Choose Your Council Members

Required Mix (check off as you identify):

- Senior leader: ____________________________
- Frontline worker: ____________________________
- Tech-savvy person: ____________________________
- Tech-hesitant person: ____________________________
- Veteran employee (10+ years): ____________________
- Newer employee (under 2 years): ____________________

Step Two: Choose Your AI Champion

Circle the person who best fits:

- Strong communicator who gets people to actually listen to them
- Well-respected across departments (not just by title)
- Genuinely excited about the company's future
- Has 2-3 hours per week available

Your AI Champion:

____

Attitude Balance

- AI enthusiast: ________
- AI skeptic: ________

Step Three: Basic Meeting Setup

Schedule: Every _____ weeks, _____ minutes

Day/Time: ____________________
Location: ____________________

Meeting Rules:
- Starts and ends on time
- Cannot be canceled for other priorities
- Send substitute if you can't attend
- Council decisions influence real company choices

Step Four: First Meeting Focus

Opening question for everyone: "What's your current AI experience and biggest question or concern?" **End with:** "What's one AI experiment we should try in the next 30 days?"

Launch Checklist:
- All 5-6 members confirmed
- AI Champion selected and briefed
- First meeting scheduled
- Leadership committed to acting on council recommendations

First Meeting Date:
Access detailed meeting templates at: **aiculturebueprint.com/aicouncil**

CHAPTER FOUR

Coach: *Shift from Mandates to Modeling*

Coaching centers on the need for organizations to empower their people to create spaces for their teams to experiment, so that new use cases and capabilities can be discovered. Too often, direction comes from the C-suite and company leadership when in reality, those folks are not on the front lines talking to customers, building products, or delivering services.

Through many in depth AI discussions within organizations over the years, I've identified five distinct personas you'll encounter who need coaching:

1. **Guardians:** Company veterans who protect quality and value experience
2. **Skeptics:** Those who need proof before adopting
3. **Innovators:** Excited early adopters at all levels
4. **Practitioners:** Frontline hands-on workflow experts
5. **Strategists:** Leaders who focus on business outcomes

Understanding how to coach each type effectively makes the difference between successful adoption and frustrated teams.

I watched this coaching failure happen with a mid-size CPA firm recently. The managing partner was frustrated because their newly formed AI Council had great ideas, but nothing was getting implemented at the team level. When I dug deeper, I discovered the problem. All the department managers were using the same approach to introduce AI tools to their teams. They'd announce the new tool in a team meeting, do a quick demo, send an email with basic steps, and expect everyone to start using it.

The Innovators jumped in immediately, the Strategists sat quietly and never touched it, and the Skeptics started complaining that *"management is forcing more technology on us."* Three months later, only 20 percent of the team was using any AI tools despite having an active, engaged AI Council generating great ideas. The missing link was that managers had no idea how to coach different types of people through AI adoption.

While strategy must start from the leadership of the entire company, it must be driven by the departments, business units, and people in the organization because they're the ones who know where the organization can benefit from AI usage.

If you have a culture of open communication, experimentation, and one that embraces failure as a path forward, that creates the energy to identify use cases. If you have a culture where leaders create unrealistic expectations that employees only work within the constructs of what those leaders believe is the best path, or employees and teammates are fearful about being discovered experimenting, it drives away innovation and the energy for people to share the remarkable things they're working on.

We can't always place blame squarely on employees and people leaders. Business leaders need to empower people leaders, middle managers, and the rest of their teams with autonomy to make those types of decisions. I see it constantly. Employees are hesitant and scared, asking, "What if we get found out?" It doesn't matter how much money you spend on AI implementation. Your AI adoption won't go anywhere if you have that fear and hesitation.

A Framework for Coaching Different Types

Everyone is different, and that remains true regarding technology. As we discussed earlier, there are five common personas you'll encounter when implementing AI in your organization. Understanding them helps you coach each type effectively instead of using the same approach for everyone and wondering why half your team isn't responding.

The High-Support Group: *Handle with Care*

The Guardians are extremely important to organizations because they're steady, understand the company, and have tremendous wisdom from years of experience as company veterans. They have the grit and work ethic to move business forward whether technology is available or not. However, their perspective will be very different when you introduce AI. They'll ask questions, show skepticism, and may have a longer adoption path because they've been conditioned to do their work without needing technology assistance.

The Skeptics are resistant to change and don't believe in outsourcing their tasks to machines. They've done their job a certain way for so long that their rigidity creates bottlenecks. Part of this is also a misunderstanding of what AI actually does. There's significant fear and job insecurity with this group. They worry that if they share ways to automate their work, they may not be seen as valuable to the company and have a large part of their identity tied into their work.

Coaching approach for both: Start with their pain points, not AI solutions. Honor their expertise. Frame AI as eliminating the boring stuff so they can focus on what they do best. Go slow and let them control the pace.

If someone says, *"I've been doing this for 20 years without machines,"* respond with *"And you've built incredible expertise. What if AI handled the administrative stuff so you could spend more time on the strategic work only you can do?"* Frame AI as enhancing their value, not replacing it.

The Empowerment Group: *Give Them Runway*

Innovators understand technology and can pick up new tools quickly. They can be conduits between business processes and AI capabilities. You must give these folks opportunities to share their knowledge, but they won't openly do it if they don't feel safe and valued.

Practitioners know the business inside and out as the frontline team. They're probably already using some type of AI without your awareness. They are most likely closest to your clients and interactive with them on the highest frequency in the organization. During AI strategy development, this group can heavily influence direction because they understand real workflows better than anyone.

Coaching approach for both: Give them autonomy to experiment. Allow them to be teachers for others. Challenge them with complex problems. Elevate their roles. Don't micromanage their exploration.

For those needing individual attention, call them by name: *"I'd love to hear what you think about this."* Give them opportunities to talk without interrupting and let them work through their thought process.

The Bridge Group: *Provide Context and Control*

Strategists jump into the driver's seat to guide company decisions usually as leaders of departments or groups of people. They may have preconceived notions about AI, but they're critical because if they're bought in, they'll drum up team support. They don't always have to know all the answers. It's okay to encourage them to say *"I don't know"* in AI contexts, as this builds trust and rapport in the process.

Coaching approach: Focus on business impact and competitive advantage. Let them set priorities and timelines. Make them heroes of successful initiatives. Connect everything back to business outcomes they care about. Ask *"How could AI help us compete better?"* and *"What would success look like from a business perspective?"*

From a human psychology perspective, everyone operates differently. Some are verbal processors who need to talk things through. Others process internally and need time to think before sharing. As you build momentum as a people leader, be mindful of different communication styles. Give folks different vehicles to share their knowledge, whether it be meetings, individual conversations, or written feedback.

For people leaders facilitating these coaching conversations, you'll be surprised at the information people are willing to share, assuming you give them appropriate platforms. Be mindful that some people simply won't share, and that's realistic. But if you can get at least half of your team engaged over three to six months, they become adoption drivers and hold peers to higher standards, so coaching doesn't have to come from company leadership or the C-suite.

Reflection Questions

1. Do you coach everyone on your team the same way, or do you adapt your approach based on their personality and experience?
2. When someone resists new technology, what's your typical response: to push harder or try to understand their concerns?
3. Can you identify which of the five personas each team member represents?
4. How do you currently handle team members who are skeptical about change?
5. What's your reaction when someone admits they tried something new and it failed?
6. When a team member tells you an AI experiment failed, what's your typical response: to focus on what went wrong or what they learned?
7. How comfortable are you with AI tools yourself? If you're not confident with AI, how does that affect your ability to coach others?

Action Tool #5:
The AI Coaching Guide

Step One: Identify Your Team's Personas

For each team member, identify their primary type:

- **Innovator** (early adopter, can teach others)
- **Guardian** (veteran, values experience, hesistant)
- **Strategist** (needs business context and control)
- **Practitioner** (knows real workflows, practical)
- **Skeptic** (resistant, fearful about change)

Name: ________ Persona: ________

Name: ________ Persona: ________

Name: ________ Persona: ________

Name: ________ Persona: ________

Name: ________ Persona: ________

Step Three: Coaching Success Methods

Track weekly for each persona:

Innovator/Practitioner: Number of colleagues helped + experiments tried
Guardian/Skeptic: Willingness to try one small experiment
Strategist: Public support + resources allocated

Step Four: Handling Common Responses

When someone says: "I don't have time to learn new tools"
Try this: "What if the tool saved you more time than it took to learn?"
When someone says: "I am worried that AI will replace me"
Try this: "What if AI handled the parts of your job you like least?"
When someone says: "I tried AI once and it was terrible"
Try this: "That's valuable feedback. What specifically didn't work?"

Step Two: Coaching Conversation Starters

For Guardians & Skeptics (High-Support Group):

- "What part of your job feels most repetitive?"
- "Where do you get frustrated with current processes?"
- "What if AI handled the boring stuff so you could focus on strategic work?"

For Innovators & Practitioners (Empowerment Group):

- "What AI experiments have you tried?"
- "Who else could benefit from what you've learned?"
- "What's the most complex problem you think AI could solve?"

For Strategists (Bridge Group):

- "How could AI help us compete better?"
- "What would success look like from a business perspective?"
- "What's our biggest competitive threat if we don't embrace this?"

Access persona-specific coaching scripts at:
aiculturebIueprint.com/coaching

CHAPTER FIVE

Cultivate: *Measure What Matters and Celebrate It*

After tremendous efforts in the first four chapters to create a culture of excitement around AI, now the focus is on playing the long game. How do we use all these tools to continue cultivating an organization that is excited about this disruptive technology while having appropriate access, expectations, and people in place so AI conversation and culture continue to thrive many years into the future as the technology evolves?

I witnessed this sustainability failure with a tech startup that had done everything right for the first year. They had executive commitment, conducted listening tours, formed an AI Council, and coached their teams through adoption. Six months in, they were at 75 percent AI tool usage, which was incredible success.

But then they stopped measuring culture metrics and celebrating wins. The AI Council meetings became sporadic and lost their way. Twelve months later when I checked back in, adoption had dropped to 30 percent. The CEO was baffled: *"We built this amazing AI culture. Why did it fall apart?"* The answer was simple: They treated AI culture like a project with an end date instead of a garden that needs continuous cultivation. Without ongoing attention, even the best AI culture withers and goes away. This is exactly why measurement and cultivation separate successful AI transformations from failed ones.

AI momentum will cease to exist if you don't track wins, losses, behaviors, and stories, whether using a scorecard or another measuring approach. These conversations may become one-time exciting discussions that are forgotten two weeks later. From a cultivation standpoint, the primary goal is to measure performance as an organization. Every organization is different and requires time to quantify what meaningful progress represents. Meaningful progress moves the organization forward without taking steps backward.

Using a scorecard to spotlight wins cannot be overstressed. The scorecard must have visibility by all at every single level of the team. It can't just be an executive dashboard if you want meaningful progress to move throughout the organization. That's how you win from an AI culture perspective: by measuring what matters, giving people visibility into what the entire organization is doing, and spotlighting wins and failures.

When the personas we discussed, especially Skeptics and Guardians, realize their peers and the company are rowing in the same direction, it's only a matter of time before they either want to join that momentum or start questioning whether they're aligned with where the organization is headed. This must happen organization-wide, tracked in some scorecard developed by the company based on what the company sells as products or services to clients.

Organizations that reach this stage experience true growth and discovery. This is when people inside companies step up to challenges and receive job titles that didn't exist twenty-four months ago. Even failures teach us the direction we need to go, but only if we're measuring and sharing them openly.

What to Measure and How

AI scorecards don’t need to be overly complicated. You don't need expensive and flashy dashboards or fancy analytics tools. Use a simple spreadsheet or document with a table. Measure what you, as a company, identify that can be quantified and that tells you whether AI culture is actually taking root.

The Four Categories of Culture Metrics

1. **Engagement Indicators:** Track how many new AI ideas emerge from team meetings and flow up to the AI Council. Monitor how many implementations you've completed in the last thirty days. Are you doing one implementation monthly? That's too slow for actual AI adoption. Or are you seeing two dozen AI implementations with 90 percent failing? That's fantastic because it shows what isn't working, allowing you to narrow focus on approaches that provide organizational value.

2. **Participation Measures:** How many AI Council meetings are you having? Not to overdo it, but frequency drives intelligence and breaks down silos. How many wins are you celebrating? Have you had two, three, or twelve? How many failures are you celebrating? Again, failures give you a sense of direction about what doesn't work.

3. **Adoption Velocity:** How many new tools have you adopted and who was the champion? Why were they the champion and how are you elevating them? How many hours of work did an AI tool save you? This demonstrates traction and momentum in the company.

4. **Culture Health:** How many bonuses or other incentives are you giving your teams because they've adopted AI?

Keep it Simple and Visible

Start with something basic. Make it centrally located and available to everybody, regardless of whether they're C-suite or intern. Use weekly tracking to see if AI adoption is going in a direction that benefits the company. When you measure and expect improved organizational efficiency, it increases margins that positively impact company valuation. These metrics help companies know whether they have funds to hire more staff or onboard more technology.

Celebrating Wins and Learning from Failures

The most important part for organizational leaders is answering, "What's in it for me?" Employees want praise for their work, a salary supporting their lifestyle, and work that makes them feel energized and alive with people they enjoy spending time with.

I've always been a fan of companies creating competition within teams to help them hold each other accountable and grow to levels they may not achieve individually. Some companies gamify this by praising the person with the most AI failures weekly, because that's intelligence. If you have seventeen unique AI failures, failure is an answer and a direction.

Companies that praise failure effectively say, "Here are seventeen ways that didn't work. That's seventeen fewer distractions for us. Congratulations on sharing that with the team." This creates visibility and excitement around adoption, spotlighting accomplishments from an AI perspective rather than hiding mistakes in corners where no one learns from them.

Most companies put failures in corners and never discuss them again. Failure should be front and center, because how can you improve as a team if you don't know what you're doing wrong? When you openly share adoption metrics, even when not trending in the right direction, you build trust. Have you had four weeks of AI failures? Keep grinding because that's four weeks of direction you've gained.

Creating Peer Accountability

You could gamify who's getting the most AI adoption department-wide and sharing the most knowledge with the AI Council. If someone determines a good way to use AI that spreads across the entire company, give them "AI User of the Week" or "AI User of the Month" titles. Spotlight and give them opportunities to lead initiatives so that when you have Skeptics or Guardians reluctant to embrace AI, these praised folks become the ones who change minds.

This natural competition creates accountability among peers, so when AI adoption happens, it doesn't come from the C-suite. It comes from people who work with each other. AI adoption becomes a talent acquisition and retention strategy because people want to work for forward-thinking, innovative companies trying new things. When they're with families during holidays or out with friends, they have things to be proud of and share. People do their best work when they're on winning and collaborating teams developing new approaches.

Adapting to Tool Evolution

AI tools are advancing at light speed. By the time you read this, some tools will be obsolete and new ones will have emerged. Organizations with strong AI cultures adapt quickly because their people aren't attached to specific tools. Instead, they're comfortable with the process of evaluating, experimenting, and adopting new approaches.

Companies that focus only on training people to use specific platforms struggle when those tools change. But companies that build culture around experimentation, learning, and adaptation thrive regardless of which AI tools become popular.

Integration into Company DNA

The most successful organizations I've worked with eventually stop talking about *"AI adoption"* because it's just become part of their culture. New employees learn AI approaches during onboarding, not in special workshops. Problems are naturally evaluated through both human and AI capability lenses. The scorecard metrics become part of regular business reviews.

When you reach this point, you've built something sustainable that will adapt and grow regardless of how AI technology evolves. Your culture becomes your competitive edge, not just your tools.

The future belongs to organizations that successfully blend human intelligence with artificial intelligence. When you complete this cultivation work, you join a small group of leaders who understand that AI transformation is fundamentally about people and not technology. Your success becomes a model for other organizations struggling with adoption. That's how we build a movement where AI conversations happen at every level, teams innovate together, and culture becomes the competitive advantage.

Reflection Questions

1. How will you keep AI culture alive when the initial excitement fades and it becomes routine?
2. Are you measuring culture health or just business outcomes from AI adoption?
3. What happens to AI momentum when your AI Champion leaves or gets busy with other priorities?
4. How often do you refresh your AI approach as new tools emerge and your team's needs evolve?
5. How does your company share wins and failures? Most companies put failures in corners and never discuss them again.
6. Are you openly sharing adoption metrics in the scorecard, centrally available and visible to all, even when not trending in the right direction? Have you had four weeks of AI failures?
7. How does your company share wins and losses across the board? What rituals and mechanisms are you using to measure that?

Action Tool #6:
The AI Culture Scorecard

Weekly Tracking

Engagement Indicators:

- New AI ideas submitted this week
- AI Council meetings held
- Cross-department collaborations started

Adoption Velocity:

- New tools/experiments tried
- Experiments scaled beyond one person
- Hours saved through AI tools

Culture Health:

- People celebrating failures/learnings
- AI conversations happening naturally
- Team members teaching others

Culture Health Check

Rate each statement: **1** = Never **2** = Rarely **3** = Sometimes **4** = Often **5** = Always

- AI conversations happen without being forced: ______
- People share failures without fear: ______
- Teams collaborate across departments on AI: ______
- Managers coach different personas effectively: ______
- Leadership actively uses and discusses AI: ______

Total Score: ______ / **25**

Action needed if score below **15**: Return to earlier chapters—culture foundation needs work.

Monthly Tracking

This Month's Wins

Person/Team:
What they tried:
Impact:

Person/Team:
What they tried:
Impact:

This Month's Learning Failures

Person/Team:
What didn't work:
What we learned:

Your Sustainability Plan

To keep momentum alive:

- Review scorecard weekly with leadership team
- Share metrics transparently with entire organization
- Adapt celebration methods as team preferences change
- Refresh AI Council membership every 3 months
- Connect AI culture progress to business results

Next review date: ________________

Access automated scorecard templates at: **aicultureblueprint.com/scorecard**

CONCLUSION

Start Today

The most important principle is that you don't need months to gain momentum around artificial intelligence. You need thirty minutes of focused leadership to drive your company forward with meaningful progress. You don't need to read all the books, listen to all the podcasts, or create the perfect implementation deck. You don't have to experiment alone. Do it together as a team.

Take action now. Don't wait six months when you feel it's safe, because you'll fall further behind by then. Momentum creation is not difficult, and you don't have to reinvent the wheel.

Your 5C Pathway Forward

1. **Commit:** Get clear on your "why." Write it down. Share it authentically. Back it with real budget allocation that shows culture matters more than technology.
2. **Communicate:** Listen to your people before you launch anything. Surface the fears, discover the hidden innovations, and understand what your team actually needs.
3. **Co-Create:** Form your AI Council with diverse voices. Let them influence real decisions. Give them an AI Champion who can drive momentum from within.

4. **Coach:** Help your people leaders coach different personas effectively. Innovators need runway, Guardians need patience, Skeptics need proof, Strategists need business context, and Practitioners need workflow integration.
5. **Cultivate:** Measure culture health, celebrate both wins and intelligent failures, and keep evolving as AI tools change. Make it sustainable, not just another initiative.

Considering the personas we've discussed, 90 percent of your AI adoption strategy already exists in your organization. You just need to know where to look and how to ask questions. The people in your company will openly share this intelligence if you give them a safe space to do so. That becomes gold for you because it drives your ability to adopt AI successfully.

Your Next Step

Start with Chapter One. Write your "why" statement this week. Share it with your team. Begin the listening tours. Form your AI Council. Everything else follows from that first commitment to building AI culture instead of just buying AI tools.

The technology will continue evolving at light speed. Your culture will be what determines whether your organization thrives or gets left behind.

Start today, start small, but most importantly, start together.

If you found this framework helpful and want to go deeper, you can find additional resources and ways to connect at aicultureblueprint.com. Whether you're looking for implementation templates, want to share your AI culture journey with other leaders, or need help facilitating these conversations in your organization, the resources are there when you're ready.

Thank you for reading this book. Onward and upwards to our future with AI.

www.ingramcontent.com/pod-product-compliance
Lightning Source LLC
Chambersburg PA
CBHW070651240926
51846CB00003BA/877
9781968247386